EVOLUTION IN HAWAII

· · · · · · · · · · · · · · · · · · ·

A SUPPLEMENT TO *TEACHING ABOUT EVOLUTION AND THE NATURE OF SCIENCE*

by Steve Olson

NATIONAL ACADEMY OF SCIEN

THE NATIONAL ACAI

THE NATIONAL ACADEMIES PRESS
Washington, D.C.
www.nap.edu

THE NATIONAL ACADEMIES PRESS   500 Fifth Street, N.W.   Washington, DC 20001

This study was supported by funds from the Council of the National Academy of Sciences. Any opinions, findings, conclusions, or recommendations expressed in this publication are those of the author(s) and do not necessarily reflect the views of the organizations or agencies that provided support for the project.

International Standard Book Number 0-309-08991-3 (Book)
International Standard Book Number 0-309-52657-4 (PDF)

Additional copies of this publication are available from the National Academies Press, 500 Fifth Street, N.W., Lockbox 285, Washington, DC 20055; (800) 624-6242 or (202) 334-3313 (in the Washington metropolitan area); Internet, www.nap.edu.

Suggested citation: National Academy of Sciences. (2004). *Evolution in Hawaii: A Supplement to Teaching About Evolution and the Nature of Science*, by Steve Olson. Washington, DC: The National Academies Press.

THE NATIONAL ACADEMIES
Advisers to the Nation on Science, Engineering, and Medicine

The **National Academy of Sciences** is a private, nonprofit, self-perpetuating society of distinguished scholars engaged in scientific and engineering research, dedicated to the furtherance of science and technology and to their use for the general welfare. Upon the authority of the charter granted to it by the Congress in 1863, the Academy has a mandate that requires it to advise the federal government on scientific and technical matters. Dr. Bruce M. Alberts is president of the National Academy of Sciences.

The **National Academy of Engineering** was established in 1964, under the charter of the National Academy of Sciences, as a parallel organization of outstanding engineers. It is autonomous in its administration and in the selection of its members, sharing with the National Academy of Sciences the responsibility for advising the federal government. The National Academy of Engineering also sponsors engineering programs aimed at meeting national needs, encourages education and research, and recognizes the superior achievements of engineers. Dr. Wm. A. Wulf is president of the National Academy of Engineering.

The **Institute of Medicine** was established in 1970 by the National Academy of Sciences to secure the services of eminent members of appropriate professions in the examination of policy matters pertaining to the health of the public. The Institute acts under the responsibility given to the National Academy of Sciences by its congressional charter to be an adviser to the federal government and, upon its own initiative, to identify issues of medical care, research, and education. Dr. Harvey V. Fineberg is president of the Institute of Medicine.

The **National Research Council** was organized by the National Academy of Sciences in 1916 to associate the broad community of science and technology with the Academy's purposes of furthering knowledge and advising the federal government. Functioning in accordance with general policies determined by the Academy, the Council has become the principal operating agency of both the National Academy of Sciences and the National Academy of Engineering in providing services to the government, the public, and the scientific and engineering communities. The Council is administered jointly by both Academies and the Institute of Medicine. Dr. Bruce M. Alberts and Dr. Wm. A. Wulf are chair and vice chair, respectively, of the National Research Council.

www.national-academies.org

Contents

Preface

As individuals and societies, we are now making decisions that will have profound consequences for future generations. How should we balance the need to preserve the earth's plants,
animals, and natural environment against other pressing concerns? Should we alter our use of
fossil fuels and other natural resources to enhance the well-being of our descendants? To what
extent should we use our new understanding of biology on a molecular level to alter the characteristics of living things, including people?

None of these decisions can be made wisely without a thorough understanding of life's history on this planet. People need to know why living things have the characteristics they do, how
those characteristics originated, and whether living things will continue to change in the future.
In short, they need to understand biological evolution.

Yet the teaching of evolution continues to be opposed on religious grounds in schools
throughout the United States. Opponents of evolution assert that the scientific justification for
evolution is lacking, when in fact the occurrence of evolution is supported by overwhelming evidence. Legislators and schools boards insert wording into laws, lesson plans, and textbooks man-
dating that evolution be taught as a controversial explanation of life's history, though no such
characterization is scientifically warranted. In some places, tremendous pressure has been exerted
on teachers and school administrators to downplay or eliminate the teaching of evolution. As a
result, many students are not being exposed to information they will need to make informed
decisions about their own lives and about our collective future.

In 1998 the National Academy of Sciences, a private, nonpartisan group of scholars that provides advice to the nation on issues involving science and engineering, released a book entitled
Teaching About Evolution and the Nature of Science. Through scientific examples, teaching exer-
cises, and dialogues among teachers, *Teaching About Evolution and the Nature of Science* summa-
rizes the observational evidence for evolution, demonstrates how the teaching of evolution can be
used to illuminate the nature of science, addresses common misconceptions about the teaching of
evolution, and offers guidance on how to choose classroom materials.

Evolution in Hawaii is a supplement to that earlier book. It examines evolution and the
nature of science by looking at a specific part of the world—the Hawaiian islands. Islands are
especially good places to see evolution in action. When plants, animals, or microbes travel from a
continent to an island, they are separated from the other members of their species and often
encounter a biological and ecological setting different from what they left behind. If the organisms that reach an island survive and produce descendants, those descendants may evolve along

different pathways than would have been the case elsewhere. By studying these evolutionary pathways, biologists have been able to draw powerful conclusions about evolution's occurrence, mechanisms, and courses.

To give students an opportunity to gain a deeper understanding of evolution, this book contains a teaching exercise similar to the ones contained in *Teaching About Evolution and the Nature of Science*. Using real genetic data from 18 species of *Drosophila* flies in Hawaii, students draw evolutionary trees depicting the relationships of the species and investigate the link between speciation and the ages of the Hawaiian islands. By letting students explore the mechanisms involved in the origin of species, the teaching exercise demonstrates how descent from a common ancestor can produce organisms with widely varying characteristics.

This publication has been designed specifically to supplement a broader consideration of evolution. By exploring a particular example in depth, it illuminates the general principles of evolutionary biology and demonstrates how ongoing research is continuing to expand our knowledge of the natural world. A related book, *Science and Creationism: A View from the National Academy of Sciences, 2nd Edition* (National Academy of Sciences, 1999), considers at greater length the arguments used by those who oppose the teaching of evolution in public schools. (*Science and Creationism* is available on the Internet at http://www.nap.edu/books/0309064066/html.)

The text of this publication was written by Steve Olson with input and assistance from Hampton Carson, Elysse Craddock, and Kenneth Kaneshiro. It was reviewed by a panel of scientists and educators that included Francisco Ayala, Wayne Carley, Gerald Carr, Brent Dalrymple, Timothy Goldsmith, Valdine McLean, Eric Meikle, Kenneth Miller, Leslie Pierce, Barbara Schulz, Rachel Wood, and Peter Vitousek. Erika Shugart and Jay Labov managed the project and contributed substantially to the development of the text. They were assisted by Dimitria Satterwhite, Kirsten Sampson Snyder, and Yvonne Wise. Additional thanks are extended to Rachel Marcus, Will Mason, Dan Parham, and Sally Stanfield at the National Academies Press for their work on the production of this book and Anne Rogers for the design and layout.

The teaching exercise was developed through the collaborative efforts of Hampton Carson and Kenneth Kaneshiro of the University of Hawaii; Elysse Craddock of Purchase College, State University of New York; LeslieAnn Pierce, Diane DeFalco, and Jay Calfee of Fairfax County Public Schools, Virginia; Steve Olson; Lyn Countryman of Malcolm Price Laboratory School in Cedar Falls, Iowa; and Judith Shaw and the students of her Advanced Placement biology class of Auburn Riverside High School, Washington.

The controversy over the teaching of evolution in the United States has been going on for many decades and will not be easily resolved. The opponents of evolution are trying in many ways to undercut evolution's place in the science curriculum. Those who are committed to the teaching of evolution have much work in front of them to continue to ensure the integrity of U.S. science education.

I have known many scientists during my life who are deeply religious. They see no contradiction between their beliefs and the teaching of evolution and are firmly opposed to introducing religious ideas into science classrooms. The scientific and the religious domains of human life are both important, but they need to remain separate if each is to contribute to a better future.

Bruce Alberts, *President*
National Academy of Sciences

E V O L U T I O N   I N   H A W A I I

. . . .

A SUPPLEMENT TO *TEACHING ABOUT
EVOLUTION AND THE NATURE OF SCIENCE*

Hawaii Is One of the Best Places in the World to Study Biological Evolution

The biodiversity of the native plants and animals that live in the Hawaiian islands is breathtaking. On the northern and eastern sides of the islands, where the trade winds drop their moisture on upsloping hills and mountainsides, luxuriant rainforests contain a profusion of species of trees, shrubs, birds, and insects. On the southern and western sides of the islands—often just a few miles away—open landscapes of grasses and shrubs receive just a few inches of rain a year. On the flanks of the high volcanoes of Maui and the Big Island of Hawaii, scattered bushes and hardy flowering plants endure frequent frosts and even an occasional snowfall. In the reefs surrounding the islands, hundreds of species of tropical fish swim amid colorful beds of coral.

Where did all these plant and animal species come from? The Hawaiian islands consist of the tops of mid-ocean volcanoes, are about 4,000 kilometers (2,500 miles) from the nearest continent, and have never been connected to any body of land (see Figure 1). Plants and animals therefore could not have taken an overland route from continents such as Asia or the Americas to reach the islands. Many of the plants and animals on the islands are so similar to species elsewhere that they obviously were brought to Hawaii by the humans who began colonizing the islands between approximately 1,200 and 1,600 years ago. But many other plant and animal species on Hawaii are so different from organisms elsewhere that they must be native—that is, they had to be present on the islands before the arrival of human beings.

The characteristics of native Hawaiian plants and animals raise further questions. Most native Hawaiian species are endemic, which means that they are found in Hawaii and nowhere else in the world. How did these unique species come to live only in the Hawaiian islands?

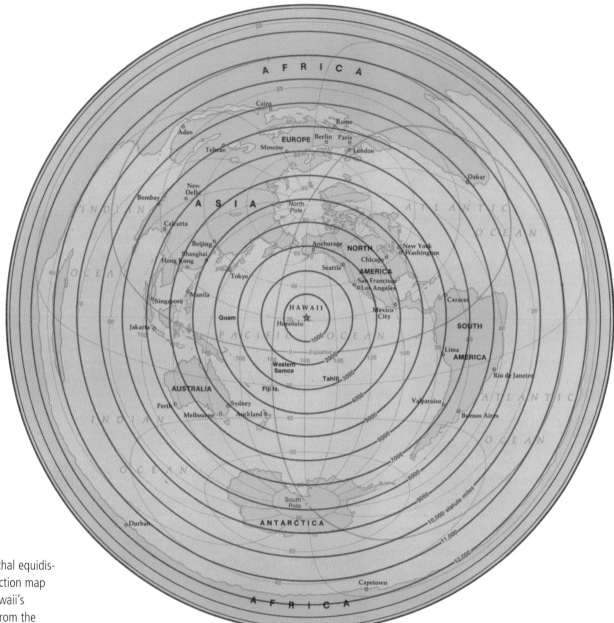

Figure 1

An azimuthal equidistant projection map shows Hawaii's isolation from the continents and other islands. Concentric circles indicate distances of 1,000 miles or about 1,600 kilometers. (Map copyright R. Warwick Armstrong, ed., *Atlas of Hawaii, 2nd Edition.* Honolulu: University of Hawaii Press, 1983.)

Furthermore, though the islands have a rich variety of flowering plants, insects, birds, land snails, and fish, they have no native species of reptiles, amphibians, or conifers, and a single species of bat and a seal species are their only living native mammals. Why do some kinds of organisms live in Hawaii but not others?

Finally, some categories of native Hawaiian plants and animals are represented with remarkable abundance. For example, approximately 800 species of flies belonging to the genera *Drosophila* and *Scaptomyza* exist in the Hawaiian islands—about a quarter of the worldwide total and far more than are found in a comparable area anywhere else on earth. But only about 15 percent of the world's total of insect families is represented in Hawaii. Before the arrival of humans, Hawaii had no native species of termites, ants, or mosquitoes, according to the fossil record. How can all these facts be explained?

Biological Evolution Explains the Characteristics of Hawaii's Plants and Animals

Only one known scientific explanation can account for the characteristics and distribution of Hawaii's flora and fauna. All of the native plants and animals in Hawaii must be descended from organisms that made their way to the initially barren islands through the air or across the water from the surrounding continents and from distant islands. In those cases where the initial colonists survived and produced descendants, the conditions existing in the Hawaiian islands resulted in the evolution of new species with traits found nowhere else in the world.

This process of biological change is described by the theory of evolution. Evolutionary theory holds that all the organisms existing on the earth today are the descendants of organisms that existed on the earth in the past. But today's organisms are not necessarily the same as past organisms, because living things have the potential to change from one generation to the next. When organisms reproduce sexually, their offspring differ from each other as well as from their parents. For example, an individual insect might have a somewhat different configuration of its wings, a plant might have a thicker stalk, or a mammal might be colored differently. Some of these new traits might be harmful for an organism, causing it to die before it can have its own offspring. But other traits might be advantageous in a particular environment. They might enable an organism to gather food more effectively, live in a place where there is less competition for

resources, or avoid predators. If an organism with an advantageous trait can more successfully exploit a particular ecological niche, it has better odds of surviving and is likely to produce more offspring than is an organism without that trait.

Many of the traits of an organism are shaped by interactions between the genetic messages encoded in its DNA and the environment in which it lives. If an individual organism has a trait encoded in its DNA that is advantageous in a given environment, its offspring can inherit that trait through the transmission of DNA between generations. These offspring can in turn pass the trait on to their offspring, so that over many generations the trait gradually becomes more common in a population of organisms. Individual organisms do not evolve during their lifetimes, but over long expanses of time successive generations of organisms can acquire new characteristics that enhance their ability to survive and reproduce.

Consider, for example, a population of drosophilid flies that lay their eggs in the rotting bark of trees in damp forests (see Figure 2). As this population expands and moves into new areas, it may encounter forests in drier regions. There some flies may attempt to lay their eggs in trees where running sap has dampened the bark. Those flies that are most able to lay their eggs in this new substrate, whether because of their behavior or the shape of their bodies, would likely produce more offspring, and these offspring will tend to inherit their parents' traits. Over time, a population of flies will develop with traits more suited to laying eggs in sap-dampened bark than in rotting bark. Indeed, just such a transition occurred several times during the evolution of the Hawaiian drosophilids.

The evolution of a new trait can have important consequences. Within a population of flies adapted to laying its eggs in sap-dampened bark, other variations in behavior or morphology can occur that will be passed on to offspring. Over time, the members of a population of flies adapted to a new environment can undergo so many changes that they no longer can, or routinely do, interbreed with flies from the parent population. At that point, the new population can be considered a new species.

The formation of new species occurs through many different mechanisms, and the study of these mechanisms remains one of the most exciting areas of modern evolution-

Figure 2

The fly *Drosophila substenoptera,* which lays its eggs in rotting bark, lives in upland forests on the island of Oahu. (Photograph courtesy of Kevin Kaneshiro.)

ary biology. Evolutionary biologists are investigating why some species have remained largely unchanged for long periods of time while others have undergone rapid diversification into multiple species. They are exploring whether evolutionary change happens gradually or in spurts separated by long periods when outward change is less dramatic. They are examining the molecular mechanisms of evolutionary change. In all of these cases, scientists are studying the details of mechanisms by which biological evolution has occurred.

Science Produces Explanations That Can Be Tested Using Empirical Evidence

Science requires that scientific explanations of phenomena be based on events or mechanisms that can be observed in the natural world. This is how science builds a base of shared observations and ideas to which new knowledge can be added.

For example, scientists studying the characteristics of plants and animals in Hawaii look for natural explanations for those characteristics. They propose hypotheses that explain the evolution of those characteristics through naturally occurring mechanisms. Then they gather additional information to test their hypotheses. Because hypotheses are based on phenomena that can be measured or observed, other scientists can test the hypotheses by gathering their own data. Based on the evidence gathered, the hypothesis can be accepted or rejected and new, more refined hypotheses can be developed.

One potential source of confusion in discussing the theory of evolution is the meaning of the words "theory," "hypothesis," and "fact." In popular usage, a "theory" is something that is not known for sure. But the word "theory" has a very different meaning in science than it does in everyday use. In science, "theory" refers to an explanation of some aspect of the natural world that is held with great confidence because it is supported by overwhelming evidence. The theory of gravitation holds that all objects are attracted to each other in proportion to their mass. The cell theory says that all living things are composed of cells.

Scientists use the word "hypothesis" to describe an idea or model that has not yet been fully tested. For example, in studying evolution in Hawaii, a scientist might hypothesize that a species on one island is descended from a species on another island. The scientist then would gather evidence to test that hypothesis. If a hypothesis is supported by the evidence, the hypothesis may contribute to more complex explanations, including theories. If the available evidence does not support a hypothesis, that hypothesis can be rejected, modified, or subjected to further testing. For a hypothesis to fall within the realm of science, it must be constructed in such a way that it potentially can be shown to be wrong—otherwise the hypothesis cannot be tested against evidence from the natural world. This demand that a hypothesis be "falsifiable" is one of the defining characteristics of scientific explanations.

A "fact," in scientific terms, is an observation that has been repeatedly confirmed by the studies of different independent scientists. In other words, it is a phenomenon that has been observed so frequently that its existence is no longer being questioned.

Because theories explain facts, they embody a greater understanding of the natural world than do observations. Without theories to explain and integrate them, facts become collections of unrelated observations. Evolutionary theory is a comprehensive explanation that integrates facts from many different areas of science. It has proven tremendously successful in explaining the basis for observed phenomena and in allowing scientists to make predictions based on existing data.

Scientific Research Has Revealed How the Hawaiian Islands Originated

The study of the origins of the Hawaiian islands provides an excellent example of how science works. For many years the geological history of the islands remained a mystery. The Big Island of Hawaii (which is sometimes called just Hawaii) forms one end of a long straight archipelago of more than 100 islands, atolls, reefs, and shoals that extend to Kure Atoll (see Figure 3), a distance of more than 2,400 kilometers (1,500 miles). Beyond Kure the chain continues underwater in the form of seamounts that rise above the ocean floor but no longer break the surface of the water. What could account for this intriguing geological formation?

In 1963 the Canadian geologist J. Tuzo Wilson proposed a hypothesis to explain the archipelago's origins. He suggested that the islands formed as the crust of the Pacific Ocean floor moved over a source of heat positioned beneath the crust (see Figure 4). This hot spot, as it came to be known, produced lava that erupted through the crust onto the ocean floor. Eventually these erupting volcanoes grew large enough to rise above sea level and form islands (see Figure 5). As the

GLOSSARY OF TERMS USED

Fact: In science, an observation that has been repeatedly confirmed.

Hypothesis: A testable statement about the natural world that can be used to build more complex inferences and explanations.

Theory: In science, a well-substantiated explanation of some aspect of the natural world that can incorporate facts, laws, inferences, and tested hypotheses.

(Adapted from *Teaching About Evolution and the Nature of Science.*)

Figure 3

The Hawaiian-Emperor volcanic chain stretches from the Big Island of Hawaii to Kure Atoll and then continues underwater as a series of seamounts. The islands currently above water are shown in solid black, with the populated chain of major islands located between Hawaii and Kauai at the bottom right. The lines that encircle the islands and seamounts indicate the 1-kilometer and 2-kilometer depth contours. (Map adapted from David A. Clague and G. Brent Dalrymple, "The Hawaiian-Emperor Volcanic Chain, Part I. Geologic Evolution," in R.W. Decker, T.L. Wright, and P.H. Stauffer, eds., *Volcanism in Hawaii*, U.S. Geological Survey Professional Paper 1350, 1987.)

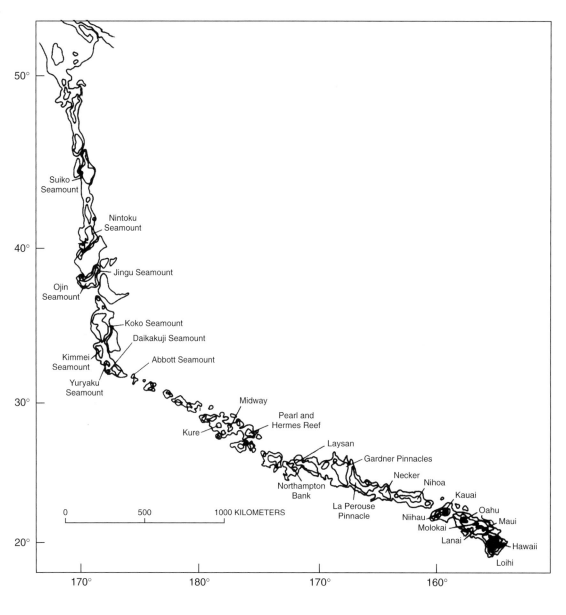

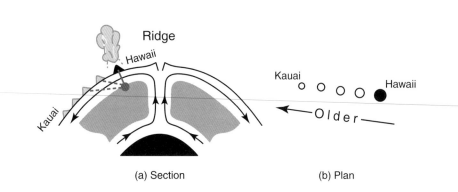

(a) Section   (b) Plan

Figure 4

In 1963 J. Tuzo Wilson proposed that the Hawaiian islands formed when the crust of the Pacific Ocean floor moved over a source of heat arising from within the earth. (Diagram adapted from J. Tuzo Wilson, "A Possible Origin of the Hawaiian Islands," *Canadian Journal of Physics* 41:863-870, 1962.)

oceanic crust moved over the hot spot, each
recently formed volcano was carried away
from the hot spot toward the northwest, cut-
ting off its source of lava. Meanwhile, a new
island was forming so that over time a chain
of islands was produced extending away from
the hot spot. As the islands continued to
move toward the northwest, away from the
hot spot, they were eroded by the wind, rain,
and waves and eventually sank below sea level
to become seamounts.

Once this hypothesis was proposed, scientists began searching for evidence to test it.
For example, the hypothesis predicts that the
more northwesterly seamounts and islands
should be older than the islands to the southeast. Geologists can measure the age of volcanic rocks by measuring the quantities of
argon gas in those rocks. Immediately after
lava cools, it contains no argon because the
gas is expelled from the molten rock. But
volcanic rocks also contain a radioactive

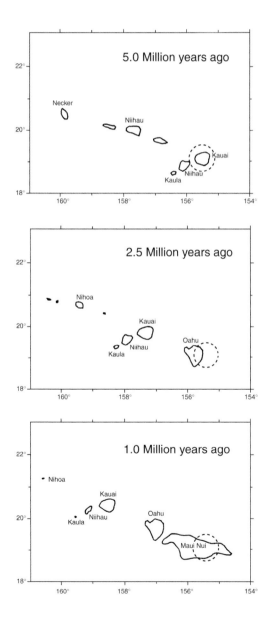

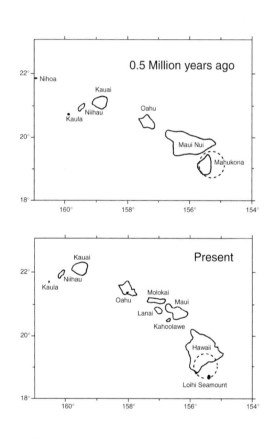

Figure 5

The Hawaiian islands formed as the Pacific Ocean floor moved over an underlying hot spot in the earth, shown here by a dotted circle. The present island of Kauai formed about 5 million years ago; Maui Nui, the landmass now represented by Maui and nearby islands, was in place more than a million years ago; and the Big Island continues to grow today. (Diagram adapted from Hampton L. Carson and David A. Clague, "Geology and Biogeography of the Hawaiian Islands," in Warren L. Wagner and V.A. Funk, eds., *Hawaiian Biogeography: Evolution on a Hot Spot Archipelago.* Washington, DC: Smithsonian Institution Press, 1995.)

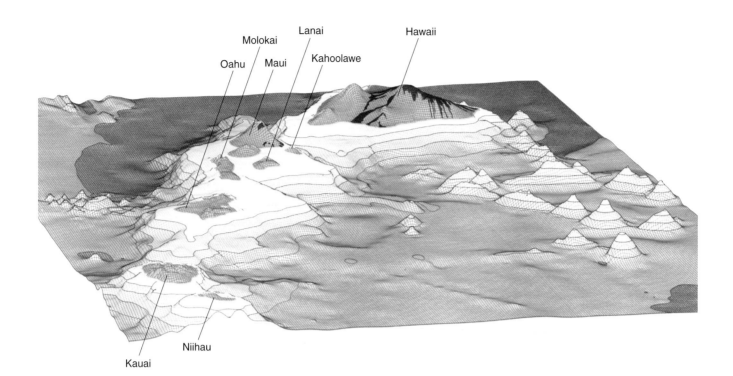

Molokai
Oahu
Maui
Lanai
Kahoolawe
Hawaii
Kauai
Niihau

Figure 6

The maximum elevations of the Hawaiian islands gradually diminish from southeast to northwest, with the newest islands being the tallest. (Map reproduced by permission of Dynamic Graphics, Inc., Alameda, CA, USA, producer of EarthVision® software, ©1984-2003 Dynamic Graphics.)

form of the common element potassium. This potassium decays at a known and constant rate into argon, and the argon remains trapped in the rock. To determine the age of a volcanic rock, scientists can measure the amount of argon and the amount of radioactive potassium in the rock. The higher the ratio, the older the rock.

These measurements showed that the Big Island of Hawaii, at the southeastern end of the archipelago, is the youngest of the chain, with an estimated age of less than half a million years (Panel 1). The islands of Maui, Molokai, Lanai, and Kahoolawe, which once were joined in a landmass known as Maui Nui, are the next older. The islands of Oahu and Kauai have greater ages, with the latter being about five million years old. To the northwest, the volcanoes are progressively older, with

Suiko Seamount in the northern part of the chain having an age of 65 million years.

This pattern is exactly what had been predicted by the hypothesis that the volcanoes were created by the movement of the crust over a source of heat. In fact, by comparing the ages of the volcanoes with their separations, geologists have concluded that the crust of the Pacific Ocean is moving at a rate of about 10 centimeters (4 inches) per year over the hot spot, or about one meter per decade.

Another source of supporting evidence involved the elevations of the islands and seamounts. As volcanic islands age, they gradually subside and erode. Thus, in Hawaii, the newest islands should be the tallest ones—which again is just what is found (see Figure 6). The highest peak on the Big Island is almost 4,250 meters (14,000 feet), while the island of Kauai rises to only about 1,500 meters (5,000 feet). The older islands beyond Kauai have much lower elevations. An example is Necker Island, which is about 300 miles northwest of Kauai and is twice as old, at about 10 million years. The geological characteristics of the underwater part of Necker Island indicate that it once was more than 1,000 meters (3,000 feet high), but today the highest point on the island is less than 100 meters (300 feet) above sea level. Beyond Kure the tops of the seamounts get progressively lower. Suiko Seamount, once an island, is now 2 kilometers (more than a mile) underwater.

A final source of supporting evidence was the observation that island building is still underway. A new underwater volcano was discovered about 32 kilometers (20 miles) off the shoreline of the Big Island (see Figure 7). Known as the Loihi Seamount, it already rises more than 3,000 meters (9,000 feet) above the floor of the Pacific and is currently within about 1,000 meters (3,000 feet) of the ocean

surface. Sometime in the next 100,000 years, it could rise above the waves to produce the newest addition to the Hawaiian islands.

Wilson's hypothesis about the origins of the Hawaiian islands, now thoroughly tested and accepted, has contributed to a much broader understanding of geological processes known as the theory of plate tectonics. About the same time that Wilson made his hypothesis, he and other geologists were recognizing that the earth's surface is broken into a number of "plates" of various sizes that can move about in relation to one another. Some plates are expanding outward from volcanic ridges that add new material to the plates' edges. Some plates are losing material as their edges plunge into the earth's interior at deep oceanic trenches. Wherever plates meet or slide against each other—at the San Andreas Fault in California, for example—earthquake activity can be particularly intense.

This theory of plate tectonics accounts for many previously unexplained geological features. For example, other hotspots beneath the

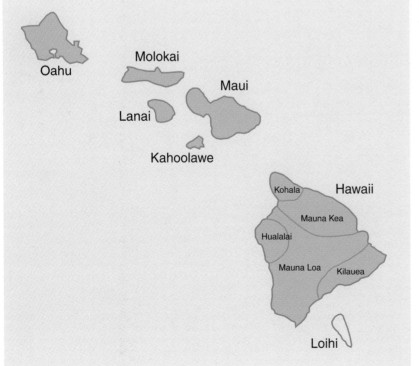

Figure 7

The Loihi Seamount off the coast of the Big Island is now above the hot spot beneath the Pacific Plate and is steadily growing. (Diagram adapted from drawing by Joel E. Robinson, U.S. Geological Survey.)

earth's crust have helped create geological formations, such as the hot spot beneath Yellowstone National Park that heats the park's hot springs and geysers. The movements of tectonic plates have had profound effects on oceanic circulation and climate—the movement of Antarctica toward the South Pole over the last 30 million years led to the formation of the southern ice cap and a gradual cooling of the planet. Similarly, the theory of plate tectonics explains many other geological features (see Figure 8). The bend in the Hawaiian archipelago past Kure Atoll, for example, appears to have resulted from a change in the direction of movement of the Pacific Plate about 43 million years ago combined with movement of the hotspot beneath the archipelago.

Figure 8

The Hawaiian islands are near the center of the Pacific Plate, which is moving toward the northwest as material is added to the plate from the midocean ridge off South America. (Diagram courtesy of U.S. Geological Survey.)

The Evolution of Life on Earth Set the Stage for Evolution in Hawaii

The history of the Hawaiian islands constitutes a very late chapter in the story of life's history on earth. The oldest member of the Hawaiian archipelago still above water, Kure Atoll, formed about 30 million years ago. The age of the earth, which has been estimated by measuring the amounts of various radioactive elements in its crust and in meteorites, is about 4.6 billion years—more than 150 times the age of Kure Atoll.

The first living things appeared on the earth surprisingly early in its history. Traces of ancient organisms have been found in rocks that are more than 3 billion years old. Biologists do not yet know how the first living

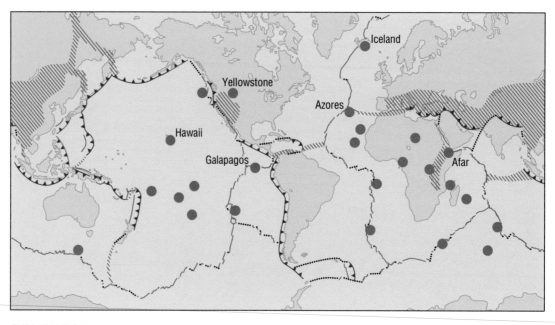

EXPLANATION

Divergent plate boundaries—
Where new crust is generated as the plates pull away from each other.

Convergent plate boundaries—
Where crust is consumed in the Earth's interior as one plate dives under another.

Transform plate boundaries—
Where crust is neither produced nor destroyed as plates slide horizontally past each other.

Plate boundary zones—Broad belts in which deformation is diffuse and boundaries are not well defined.

 Selected prominent hot spots

things formed. But they have identified and replicated several chemical processes that could have created the complex carbon-based molecules necessary for life, and they are investigating hypotheses about how these molecules could have come together in assemblages that could make copies of themselves.

Evolution must have been a part of life from the very beginning. The mechanisms that copy the genetic material in cells are not 100 percent accurate. Thus, as the first primitive cells multiplied, they would have begun to produce variants, some of which would have had advantages over their predecessors. Expanding populations of cells also would encounter new environments, which would favor variants that could survive and produce more offspring in the new conditions. In this way, living things diversified over time and took advantage of a growing number of different ecological niches.

Over the billions of years of earth's history, several important evolutionary milestones have occurred:

- The appearance of eukaryotic cells more than a billion years ago resulted in organisms with traits not found in earlier prokaryotic cells, since the development of cellular organelles (such as a membrane-bound nuclear envelope) enabled cells to be larger and more complex.

- Similarly, the evolution of multicellular organisms with cells specialized for particular tasks gave rise to new kinds of organisms with novel forms and functions.

- The ability to grow hard shells, which appears to have evolved about 570 million years ago, led to the appearance of a variety of new species of animals, with some lines of organisms flourishing and others dying out.

- The evolution of plants and animals that could live exclusively on land, beginning about 400 million years ago, led to a further diversification of forms and functions.

Vascular plants, for example, developed true leaves that enabled more efficient photosynthesis, stems to support the leaves, and roots to anchor the plant as well as to absorb water and nutrients from the soil.

- The first reptiles, birds, and mammals evolved between about 300 million and 200 million years ago, so that for the first time in earth's history the surface of the planet was occupied by a wide range of large terrestrial animals.

- The extinction of the dinosaurs about 65 million years ago enabled mammalian species to expand into a range of ecological niches from which they were previously excluded.

About 30 million years ago, as what is now Kure Atoll was being formed, the world was acquiring many of the characteristics we recognize today. The continent of Antarctica had separated from South America and was moving toward its present location. The resultant cooling of worldwide temperatures produced a new type of vegetative zone in higher latitudes, a temperate mixed woodland similar to the forests of Canada and northern Europe today. Species with distinct similarities to modern dogs, cats, camels, pigs, and deer appeared about this time. In the northern latitudes, many species of primates went extinct as the climate cooled, but in the tropics several evolutionary lineages of primates survived. Millions of years later one of these lines would give rise to gorillas, chimpanzees, and human beings.

When each new Hawaiian island rose above the waters of the Pacific, it was as hot and lifeless as the surface of the early earth (see Figure 9). But it did not remain barren for long. As soon as the lava cooled, it was colonized by spores of algae, mosses, and ferns carried on the wind from other islands and from distant continents. In addition, some birds, bats, and insect species are capable of

Panel 1

PLATE TECTONICS AND THE AGES OF THE HAWAIIAN ISLANDS

Many observers of the Hawaiian islands, including the original Polynesian colonists, have concluded that the northwestern islands are older than the southeastern ones because they are lower, more eroded, and no longer volcanically active. According to Hawaiian legend, for example, the goddess Pele moved successively from the older islands to the newer ones, creating new volcanoes on each island in an effort to escape the wrath of her sister Namaka.

With the development of potassium-argon radioactive dating in the 1960s, geologists were able to determine the actual ages of each Hawaiian island. For example, one comprehensive set of measurements has produced the following ages for the volcanoes on different islands:

ISLAND	VOLCANO	AGE (millions of years)
Hawaii	Mauna Kea	0.375
Hawaii	Kohala	0.43
Maui	Haleakala	0.75
Maui	West Maui	1.32
Lanai	Lanai	1.28
Molokai	East Molokai	1.76
Molokai	West Molokai	1.90
Oahu	Waianae	3.7
Niihau	Niihau	4.9
Kauai	Kauai	5.1
Nihoa	Nihoa	7.2
Necker	Necker	10.3

SOURCE: David A. Clague and G. Brent Dalrymple. The Hawaiian-emperor volcanic chain, Part I. Geologic evolution. In R.W. Decker, T.L. Wright, and P.H. Stauffer, eds. *Volcanism in Hawaii*, U.S. Geological Survey Professional Paper 1350:5-54, 1987.

If the ages of the islands are plotted against their distance from the currently erupting volcano on the Big Island of Hawaii, the following graph results:

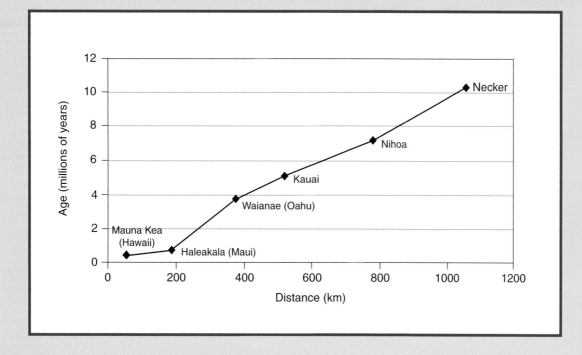

This graph supports the hypothesis that the Hawaiian archipelago formed as the Pacific Plate moved at a relatively constant rate over a hotspot that is currently located under the Big Island. As the plate moved to the northwest, islands moved away from the hotspot and began to erode and subside while new islands were created over the hotspot.

This graph also can be used to calculate the rate at which the Pacific Plate is moving. Necker Island is now about 1,050 kilometers (650 miles) from the Big Island, so the Pacific Plate must have moved that distance over the past 10.3 million years. This rate of movement is equivalent to about 10 centimeters (4 inches) per year.

flying the thousands of miles to the islands
from Asia or the Americas, especially if aided
by strong winds. Birds, in turn, often carry
seeds and other organisms in their guts or
stuck to their feathers, beaks, and feet. Insects,
spiders, snails, and other small organisms likely rafted to the islands on floating branches or
mats of vegetation. And fishes, mollusks, sea-
weeds, and other marine organisms found
new homes on the underwater flanks of the
volcanoes after swimming to the islands or
being carried there by oceanic currents.

Thus each new Hawaiian island was colonized by a variety of plant and animal
species. But because of the islands' isolation
in the middle of the Pacific, only a small
fraction of the species from surrounding
landmasses likely reached Hawaii. For
example, about 2,500 species of bony fishes
live in the near-shore waters of the
Philippines, but only about 530 occupy
Hawaiian waters. Only a single genus of
palm, the loulu palm, became established in
Hawaii before the arrival of humans, though
up to 100 genera of the family occur on
other islands in the southwestern Pacific.
And only 6 of 174 families of songbirds
worldwide are native to Hawaii.

Once a newly introduced species became
established in the Hawaiian islands, it could
remain part of a widely distributed species
found both there and elsewhere. For example,
many of the fish species that live in Hawaii
receive continued immigrants from surrounding regions and remain genetically linked to
species distributed throughout the Pacific.

Alternately, a newly established species in
Hawaii could evolve into one or more new
species. In some cases, this resulted in just a few
new species. For example, several known species
of flightless ducks, all now extinct, appear to be
descended from a single duck species that
colonized the islands, probably from North
America. In other cases, the conditions encoun-
tered by colonizing species led to an explosive
proliferation of new species, as demonstrated by
the flies known as drosophilids.

An Adaptive Radiation Has Led to a Dramatic Diversification of the Drosophilids in Hawaii

In an area of just 16,700 square kilometers
(about 6,500 square miles), the Hawaiian
islands have the most diverse collection of
drosophilid flies found anywhere in the world
(see Figure 10). Different species range in
body length from less than 1.5 millimeters
(a sixteenth of an inch) to more than 20 millimeters (three-quarters of an inch). Their
heads, forelegs, wings, and mouthparts have
very different appearances and functions.
Hawaiian drosophilids live everywhere from
sea-level rainforests to subalpine meadows.
Some species produce one egg at a time while
others produce hundreds.

The approximately 800 native drosophilid
species in Hawaii belong to two genera—
Drosophila and *Scaptomyza*—which in turn are
part of the family Drosophilidae. *Drosophila*
and *Scaptomyza* are two of approximately
10,000 genera in the order Diptera, which
includes flies, gnats, and mosquitoes. It is a
tremendously diverse and successful group of

Figure 10

Three species of Hawaiian drosophilids—*Drosophila planitibia*, *D. grimshawi*, and *D. cilifera*—exhibit some of the morphological variety found in the picture-winged group of Hawaiian *Drosophila*. *D. planitibia*, with a body size of as much as 8 millimeters (0.3 inches) is almost twice the size of the other two species shown here. (Photographs courtesy of Kevin Kaneshiro.)

organisms: the fly species on earth far outnumber all of the vertebrate species combined. But the native insects of the Hawaiian islands include very few separate fly genera, and most of the native fly species are drosophilids.

When biologists began to study the evolutionary history of the Hawaiian drosophilids, they first examined the physical similarities and differences of the species. If two species have very similar appearances, scientists might hypothesize that both are descended from an ancestral species that lived quite recently. If two species are physically quite distinct, scientists could infer that they are more distantly related. Researchers then would seek additional evidence to support or reject these hypotheses. For example, two species can develop similar adaptations if they live in similar environments and therefore can appear to be more closely related than they actually are.

In recent decades, biologists have gained an additional way of examining the relationships among species. Each individual fly has a particular sequence of the chemical units that make up the DNA in its cells. In general, these sequences are more similar among the members of a single species than they are between the members of different species. Similarly, DNA sequences generally are more similar between closely related species than they are between more distantly related species. Genetic sequences accumulate changes over the generations as DNA randomly mutates and is influenced by natural selection or other evolutionary processes. If the DNA sequences of two *Drosophila* species are more similar, the two species are more likely to be descended from a relatively recent ancestral species, because their DNA has not had much time to diverge. If the DNA sequences are less similar, the two species had more time to accumulate genetic changes, indicating that their common ancestral species lived in the more distant past.

Study of the physical and genetic differences among the hundreds of species of native drosophilids in Hawaii has led scientists to a remarkable conclusion. All of the native *Drosophila* and *Scaptomyza* species in Hawaii appear to be descended from a single ancestral species that colonized the islands millions of years ago! In fact, all of the approximately 800 species of drosophilids in Hawaii could be descended from a single fertilized fly that somehow reached the islands—perhaps blown there by a storm, or carried to the islands in a scrap of fruit stuck to the feathers of a bird.

Since that time, the descendents of the original colonists have undergone what evolutionary biologists call an adaptive radiation. New species have evolved and have occupied a wide range of ecological niches. Several interacting factors have contributed to this adaptive radiation. An especially important factor for the Hawaiian drosophilids has been what is called the founder effect. Many new populations of drosophilids in Hawaii must have become established in much the same way as did the original population. A few individuals or a single fertilized female must have journeyed or been transported from one area of suitable habitat within an island to another

such area, or from one island to another. These founders carried with them just a subset of the total genetic variability within its species. As a result, the physical characteristics and behaviors of the founders could differ from those typical of the parental population. Under such circumstances, a founder population can diverge from the ancestral population and eventually may become a new species.

The great ecological diversity of the Hawaiian islands also plays a role in adaptive radiations. *Drosophila* species continually expanded into wetter or drier areas, higher or lower elevations, and regions of differing vegetation. The members of a species able to survive in these new areas can acquire new adaptations that set them apart from the original species.

Finally, the lack of competitors in island settings can spur the evolution of new species. In Hawaii, the drosophilids could move to new islands or into ecological niches that on the continents would already have been filled by other species. For example, many Hawaiian drosophilids lay eggs in decaying leaves on the ground, an ecological niche that

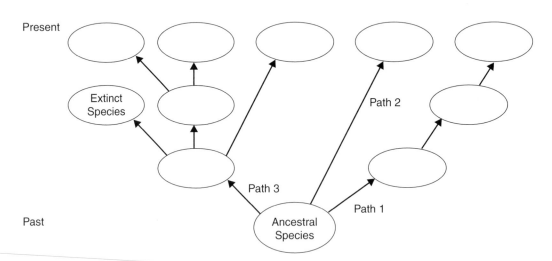

Figure 11

An ancestral species that lived in the past can give rise to multiple species through a variety of different evolutionary pathways. An ancestral species can gradually evolve into a series of what would be considered new species while remaining a single genetically connected population (path 1). Or a species can remain unchanged for a long period of time (path 2). Or an ancestral species can undergo a series of splits, generating new species that in turn become extinct or undergo further speciation (path 3).

Figure 12a

Two males of the species *D. silvestris* grapple head to head. (Photograph courtesy of Kevin Kaneshiro.)

Figure 12b

Map of the Big Island shows five regions inhabited by populations belonging to the species *D. silvestris* (shaded in gray). The populations on the Hilo side of the island, which are younger than the Kona side populations, are developing differences that over time could lead to the origin of a new species. (Map adapted from Hampton L. Carson, "Sexual Selection: A Driver of Genetic Change in Hawaiian *Drosophila*," *Journal of Heredity* 88:343-352, 1997. The contours are in meters.)

is filled by many organisms on the continents but in early Hawaii was almost empty.

As one species diversifies into many, a variety of different evolutionary paths can be taken (see Figure 11). An ancestral species can give rise to a daughter species while remaining relatively unchanged itself. Or a succession of single species can lead from an ancestral species to a single current species. Or an ancestral species can undergo repeated divisions, producing complex networks of evolutionary relationships (Panel 2).

The speciation of drosophilid flies in Hawaii is continuing to occur. For example, a species known as *Drosophila silvestris* occupies several discrete patches of forest on the Big Island (see Figures 12a and 12b), living in cool, wet forests above 750 meters (2,500 feet) in elevation and laying its eggs in the decaying bark of trees. Males of *D. silvestris* have a series of hairs on their forelegs that they brush against females during courtship. On the northeastern half of the island (known as the Hilo side), the males have many more of these hairs than do the males on the southwestern side (the Kona side). These two populations are developing physical and behavioral differences that over time might split a single species into separate species.

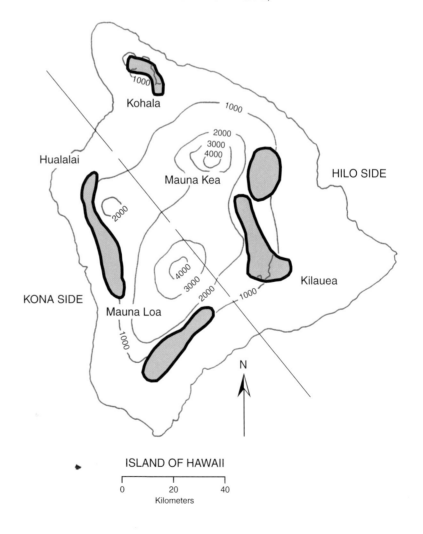

Panel 2

EVOLUTIONARY RELATIONSHIPS AMONG THE PICTURE-WINGED DROSOPHILIDS

About 100 species of Hawaiian *Drosophila*, such as the male *D. nigribasis* shown here, have large patterned wings and are known as picture-winged *Drosophila*. (Photograph courtesy of Kevin Kaneshiro.)

The drosophilid flies of Hawaii offer one of the best examples in the world of an adaptive radiation, where an ancestral species has given rise to hundreds of new species. In particular, evolutionary biologists have focused their attention on a group of about 100 drosophilid species that have characteristic light and dark markings on their large wings. Known as the picture-winged drosophilids, these species carry within them a remarkable biological record of the group's evolutionary history.

Drosophila has long been a favorite model organism for geneticists, in part because the chromosomes in some of their cells condense into large, thick structures that are easily visible through a microscope. Biologists first capture the flies in their natural habitats and bring them into the laboratory. There the female flies lay eggs that are grown to the larval stage. At that point, the salivary glands of the larvae are extracted, and chromosomes from the salivary cells are stained and mounted on slides. When viewed with a microscope, these so-called polytene chromsomes display hundreds of alternating dark and light bands of differing sizes.

Chromosome from one *Drosophila* species shows break points delimiting a section of DNA that is inverted in some other species. In chromosomes with inversions, the marked section occurs in a reversed orientation. (Photograph courtesy of Hampton L. Carson, based on an original photograph by Harrison D. Stalker, Washington University, St. Louis.)

Using polytene chromosomes, it is especially easy to detect a kind of chromosomal rearrangement called an inversion. Sometimes a chromosome can be damaged at two separate places. When molecular mechanisms within the cell seek to repair the damage, a segment of the chromosome can be flipped with no effect on the functioning of the chromosome. The result is a rearranged chromosome in which a section of the chromosome, with its characteristic light and dark bands, has a reversed orientation.

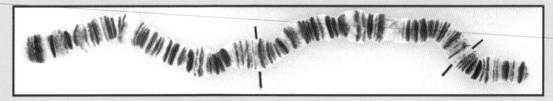

Many such mutations are lost when a chromosome with a new inversion is not passed on to offspring. But in other cases many descendants of the fly in which the mutation first occurred inherit the chromosome with the inversion, and in some cases the structurally modified chromosome will spread with each new generation until it is found in all the members of a given species. In this way, different *Drosophila* species come to have different genetic banding patterns. Their chromosomes thus act as a sort of genetic bar code, allowing separate species to be compared.

Beginning with the work of University of Hawaii geneticist Hampton Carson in the 1960s, researchers have used these banding patterns to work out the evolutionary history of the picture-winged drosophilids. If two species are descended from a single ancestral species, they will each inherit the bar code of their ancestor. At the same time, one of the daughter species may carry a new inversion that distinguishes it from its sister species, and it will pass this new marker on to any species that descend from it.

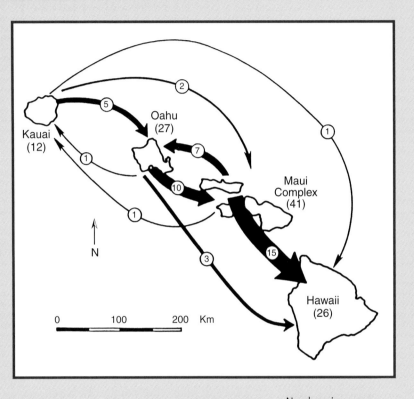

By studying these chromosomal banding patterns, biologists have reconstructed the sequence in which existing species of drosophilids moved from older islands to newer islands and formed new species. For example, the Big Island of Hawaii currently has 26 species of picture-winged drosophilids. Most of these species resulted from 15 founder events from the Maui complex of islands. In each of these events, a small group of flies (or perhaps a single fertilized female fly) journeyed from Maui or its surrounding islands to the Big Island. These founder events produced one or more drosophilid species that became distinct from the ancestral species they left behind. The remaining picture-winged species of the Big Island appear to result from three founder events from Oahu and one from Kauai. All of these species on the Big Island formed within the past half million years and are found nowhere else in the world.

Numbers in arrows indicate the number of founder events proposed to explain the origin of the picture-winged drosophilids on each island group. The number of picture-winged species found on each island is given in parentheses. (Diagram adapted from Hampton L. Carson, "Chromosomal Sequences and Inter-Island Colonizations in Hawaiian *Drosophila*," *Genetics* 103:465-482, 1983.)

Many Other Species Have Undergone Adaptive Radiations in Hawaii

While the adaptive radiation of the drosophilids in Hawaii has resulted in a remarkable number of species, other radiations have produced descendants with an even greater range of physical characteristics. The 30 species of plants belonging to what is called the silversword alliance are a superb example.

For decades, botanists had known that all of these species are related because their leaves and flowers share certain characteristics, which is why the species are termed an "alliance." But in other ways the plants are so different that the nature of this relationship remained obscure. The 30 members of the alliance include trees, shrubs, mats, vines, and the rare and magnificent flowering silverswords that live on the high slopes of Haleakala in Maui and Mauna Kea in the Big Island (see Figure 13). They occupy habitats ranging from near sea level to the upper limits of vegetation on mountains and from semi-arid desert areas to thick rainforests.

In the 1980s, biologists studying the genetics of these species realized that they are far more closely related than they appear. In fact, all appear to be descended from a single ancestral species that arrived on the Hawaiian islands millions of years ago (Panel 3). Today the closest non-Hawaiian relative of the silversword alliance is a small, daisy-like plant called a tarweed that grows on the west coast of North America. The fruit of this plant has sticky appendages that could easily have allowed the seeds of an ancestral plant to hitch a ride on a migratory bird.

As with the drosophilids, the descendents of the original colonizing species diversified dramatically as they underwent multiple founder events and spread into new environments. On Maui and the Big Island, for example, several species within the silversword alliance have adapted to the drier environments of higher elevations. All of these species are able to maintain water pressure within their leaves and stems even as the water content of the plant drops, which helps them survive in drier habitats.

Many other species of plants and animals in Hawaii also derive from adaptive radiations.

- About 50 living and extinct species of the birds known as honeycreepers have been identified in Hawaii. All evolved from a single finch-like colonist species with a relatively small bill. Today, the members of this radiation have a wide variety of bill shapes that are each specialized for a particular kind of food (see Figure 14).

- About 240 species of crickets have evolved from the separate arrival in Hawaii of a tree cricket, a sword-tail cricket, and a ground cricket. Among these species are several species that have adapted to subterranean life within underground lava tubes in Maui and the Big Island (see Figure 15). These species have reduced coloration, small eyes, and a clear exoskeleton.

- Two genera of violets—*Viola* with seven species, and *Isodendrion* with four—derive from two separate introductions, each of which was followed by a moderate degree of adaptive radiation.

Altogether, the approximately 1,700 species of native Hawaiian plants are descended from about 300 separate species that colonized the islands. The 10,000 species of native Hawaiian insects may be descended from only 350 to 400 separate founders. Biologists are now studying the different patterns of adaptive radiations seen in Hawaii to better understand the factors that influence evolutionary diversification.

Figure 13

Descendants of a single ancestral plant species have evolved into many different forms in Hawaii, collectively called the silversword alliance. Counterclockwise from above, a vine, mat, shrub, tree, and the silversword plant. (Photographs courtesy of Gerald Carr.)

Panel 3

THE EVOLUTIONARY HISTORY OF THE SILVERSWORD ALLIANCE

High on the Hawaiian volcanoes of Haleakala and Mauna Kea is a remarkable plant that attracts a steady stream of admirers from around the world. *Argyroxiphium sandwicense,* commonly known as the silversword, is for most of its life a globe-shaped plant with rigid leaves covered by dense silver hairs. After living for anywhere from 15 to 50 years, the plant produces a spectacular stalk up to six feet tall covered with hundreds of maroon flowers; soon after producing a new generation of seeds, it dies. Near extinction in the 1920s because of destruction by humans, goats, and cattle, the silversword is now protected and is gradually increasing in numbers.

The silverswords that grow on the flanks of Haleakala and Mauna Kea represent one of five separate silversword species found on Maui and the Big Island (one of these species is now extinct). Recognizing similarities in their leaves and flowers, botanists have known for decades that these five species are related to 25 other Hawaiian plant species. Two of these other species belong to the genus *Wilkesia* and grow only on the island of Kauai. The other 23 species belong to the genus *Dubautia* and grow from Kauai to the Big Island.

In the late 1980s, botanists Bruce Baldwin, Donald Kyhos, and Gerald Carr, after comparing the chromosomes and DNA of the silversword species with that of various California species, succeeded in producing vigorous hybrids between the Hawaiian plants and a California tarweed, demonstrating that they are closely related. Further genetic research has demonstrated that all the members of the silversword alliance are descended from a single ancestral species that colonized the Hawaiian islands. Since then, biologists have used the genetic differences within the alliance to derive an evolutionary phylogeny showing how the species are related. These data suggest that there was a single species from which all the current members of the alliance evolved and that this species lived on what is now the island of Kauai about five million years ago.

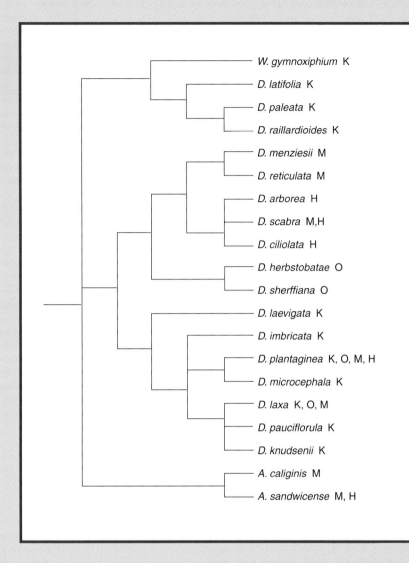

An evolutionary tree based on DNA sequences of silversword alliance members shows how current species are descended from a common ancestral species. Only some of the known species are shown here. Letters following each species indicate the island or islands on which the species grows today (K = Kauai; O = Oahu; M = Maui, Lanai, and Molokai; and H = the Big Island of Hawaii). According to the DNA data, the ancestral species split into two separate species approximately five million years ago. One of these species eventually gave rise to the *Argyroxiphium* species that now grow on Maui and the Big Island. The other species evolved into the two *Wilkesia* species that today are found on Kauai and the 23 *Dubautia* species found on all the islands. (Diagram adapted from Bruce G. Baldwin and Michael J. Sanderson, "Age and Rate of Diversification of the Hawaiian Silversword Alliance (Compositae)," *Proceedings of the National Academy of Sciences* 95:9402-9406, 1998.)

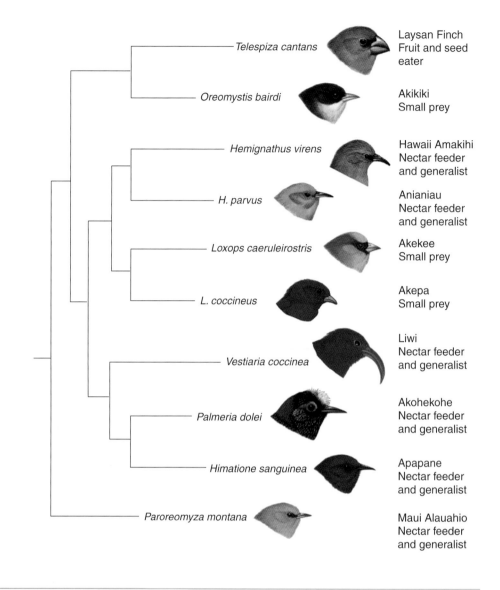

Figure 14

Species of Hawaiian honeycreepers descended from a common ancestor, some of which are shown here, have evolved many different bill shapes and food sources. (Paintings copyright H. Douglas Pratt, *The Hawaiian Honeycreepers: Drepanidinae.* Oxford: Oxford University Press, 2003. Diagram adapted from T.J. Givnish and K.J. Sytsma, eds., *Molecular Evolution and Adaptive Radiation.* Cambridge: Cambridge University Press, 1997.)

Telespiza cantans — Laysan Finch, Fruit and seed eater

Oreomystis bairdi — Akikiki, Small prey

Hemignathus virens — Hawaii Amakihi, Nectar feeder and generalist

H. parvus — Anianiau, Nectar feeder and generalist

Loxops caeruleirostris — Akekee, Small prey

L. coccineus — Akepa, Small prey

Vestiaria coccinea — Liwi, Nectar feeder and generalist

Palmeria dolei — Akohekohe, Nectar feeder and generalist

Himatione sanguinea — Apapane, Nectar feeder and generalist

Paroreomyza montana — Maui Alauahio, Nectar feeder and generalist

Figure 15

Species of crickets that live underground in Hawaii have reduced eyes and pale color compared to the surface-dwelling species that colonized the islands. (Photograph courtesy of William P. Mull.)

Alien Species Pose a Severe Threat to Hawaii's Native Plants and Animals

Figure 16
The banana poka vine, introduced into Hawaii in the 1920s as an ornamental plant, has spread throughout Hawaii's forests and fields, where it overgrows and smothers native plants. (Photographs courtesy of Gerald Carr.)

The colonization of the Hawaiian islands by humans has had a dramatic impact on the native plant and animal life of the archipelago. When the Polynesians arrived on the islands 1,200 to 1,600 years ago, they brought with them a number of plant and animal species, including taro, sugarcane, coconut palm, pigs, and chickens. The arrival of European colonizers over the past 200 years greatly accelerated the rate of introduction of alien species. Today almost as many non-native plants as native plants grow wild on the islands. Hundreds of alien species of insects, birds, mammals, and other animals have gained firm footholds. Some of these introductions were deliberate, as when crop plants, ornamental flowering plants, and non-native songbirds were brought to the islands (see Figure 16). Others were accidental, as when alien insects and animals traveled to Hawaii in cargo, ship holds, or even the wheel wells of airplanes.

Many introduced species have had little or no effect on the native flora and fauna of the Hawaiian islands. But the impact of some alien species has been devastating. Introduced goats, pigs, sheep, and cattle graze on native plants. Introduced birds, feral cats, mongooses, and avian malaria have been eliminating native birds. The Formosan ground termite now causes many millions of dollars of damage to structures in Hawaii each year. Yellowjackets (see Figure 17) introduced into the islands in a shipment of Christmas trees have been eliminating insect populations, which in turn reduces food supplies for native birds. Land clearing, agriculture, and urbanization have transformed lowland ecosystems. Since

humans arrived in Hawaii, at least 10 percent of the native Hawaiian plants have gone extinct, and an additional 40 to 50 percent are threatened or endangered.

Evolutionary theory helps explain why many native species have been so vulnerable to invaders in Hawaii. Before humans began introducing non-native animals and plants in great numbers, fewer natural predators lived in Hawaii than on the mainland. As a result, many native species gradually lost preexisting defenses, such as thorns or toxins, because variants without the defenses might have enjoyed one or more adaptive advantages. (For example, they might have been able to use for other purposes the metabolic energy that would have been expended to build defenses that were not needed in the environment of Hawaii.) Many native Hawaiian species also developed adaptations allowing them to move into ecological niches that they could not have occupied on the mainland. When alien species were introduced into the islands, they had competitive advantages that many native species did not.

An understanding of evolution is essential when anticipating and mitigating the possible effects of alien species on an ecosystem. The processes of evolution contribute to the

Figure 17

The spread of introduced yellowjackets in Hawaii since the early 1980s has decimated many native insect species. (Photograph courtesy of David Foote, Pacific Island Ecosystems Research Center, U.S. Geological Survey.)

assemblage of species in a given environment. The introduction of new species into that environment therefore will have consequences that reflect the evolutionary history of the ecosystem. By understanding the role of evolution in shaping an ecosystem, it may be possible to reduce the impact of introduced species. For example, modern ecosystem management techniques subject entire ecosystems to regular treatments, such as simulated natural fire regimes, that favor native species over introduced species.

Many Critical Problems Cannot Be Addressed Without Understanding Biological Evolution

Responding to the impact of introduced species, in Hawaii and elsewhere, is just one area where evolutionary theory can affect public policy. Many other examples can be cited. Many pathogenic microorganisms are evolving resistance to the drugs that have been used to control them, leading to outbreaks of disease. The AIDS virus changes its genetic makeup so quickly within its human hosts that the drugs that have been used to counter the virus gradually lose their effectiveness. Agricultural pests have evolved defenses against pesticides, and based on our understanding of evolution they will continue to do so.

The only way to counter these threats effectively is to anticipate the ability of populations of living things to change over time. For example, evolutionary theory points toward strategies that can greatly reduce the development of antibiotic resistance, such as reducing the overall use of antibiotics by humans and curtailing the use of antibiotics in agricultural animals eaten as food.

An understanding of evolution is essential for another reason. As we learn more about the natural world, we will have greater potential deliberately to intervene in that world. Researchers already have begun to make specific changes in the DNA sequences of plants and animals to increase agricultural productivity. Soon they are likely to gain a comparable ability to change the DNA sequences of human cells. Wise decisions about when and how to use such knowledge will require a sophisticated grasp of evolutionary theory.

Apart from its practical applications, the theory of evolution has proven to be one of the most important unifying ideas in modern science. The basic mechanism at the heart of evolution—the differential reproductive success of organisms with different heritable traits—has profound implications throughout all of biology. Evolutionary theory is even central to such challenges as understanding the origins of humans and the evolution of the human brain. That such a simple mechanism could produce such complex outcomes provides a key insight into the workings of biological systems.

The flora and fauna of the Hawaiian islands offer just one example of how the processes of evolution have shaped life on earth. As such, the history of evolution in Hawaii demonstrates how powerful the concept can be in explaining what we see around us. The study of biological evolution has produced knowledge that is not only immensely useful but profoundly beautiful. It has deepened and enriched our view of living things—whether considering the plants and animals of Hawaii or the more than three billion years of life's history on this planet. As Charles Darwin wrote in the final sentence of *On the Origin of Species,* "There is a grandeur in this view of life, with its several powers, having been breathed into a few forms or into one; and that, whilst this planet has gone cycling on according to the fixed law of gravity, from so simple a beginning endless forms most beautiful and most wonderful have been, and are being, evolved."

Selected Bibliography

Carson, Hampton L. (1997). Sexual selection: A driver of genetic change in Hawaiian *Drosophila. Journal of Heredity* 88:343-352.
An eminent evolutionary biologist looks back on more than three decades of work with Hawaiian drosophilids.

Carson, Hampton L. (1999). Selection, Darwinian fitness and evolution in local populations. In S.P. Wasser, ed., *Evolutionary Theory and Processes: Modern Perspectives.* New York: Kluwer.
An examination of the meaning of fitness within the broader context of evolutionary theory.

Clague, David A., and G. Brent Dalrymple. (1987). The Hawaiian-Emperor volcanic chain, Part I. Geologic evolution. In R.W. Decker, T.L. Wright, and P.H. Stauffer, eds., *Volcanism in Hawaii,* U.S. Geological Survey Professional Paper 1350:5-54.
A thorough description of the geology of the Hawaiian archipelago.

Cox, George. (1999). *Alien Species in North America and Hawaii.* Washington, DC: Island Press.
An analysis of the effects of non-native species on North American and Hawaiian terrestrial freshwater, and marine ecosystems.

Craddock, Elysse M. (2000). Speciation processes in the adaptive radiation of Hawaiian plants and animals. *Evolutionary Biology* 31:1-43.
An accessible and authoritative overview of the evolutionary processes that have led to high levels of speciation in Hawaii.

Kambysellis, Michael P., and Elysse M. Craddock. (1997). Ecological and reproductive shifts in the diversification of the endemic Hawaiian *Drosophila.* In T.J. Givnish and K.J. Sytsma, eds., *Molecular Evolution and Adaptive Radiation.* Cambridge: Cambridge University Press.
An analysis of evolutionary mechanisms involved in the adaptive radiation of the drosophilids in Hawaii.

Kaneshiro, Kenneth Y. (1989). The dynamics of sexual selection and founder effects in species formation. In L.V. Giddings, K.Y. Kaneshiro, and W.W. Anderson, eds., *Genetics, Speciation, and the Founder Principle.* New York: Oxford University Press.
A study of how speciation among Hawaiian drosophilids illustrates principles involved in speciation in general.

Kaneshiro, Kenneth Y. (1993). Introduction, colonization, and establishment of exotic insect populations: Fruit flies in Hawaii and California. *American Entomologist* 39:23-29.
The problem of alien species in an evolutionary perspective.

Wagner, Warren L., and V.A. Funk, eds. (1995). *Hawaiiian Biogeography: Evolution on a Hot Spot Archipelago.* Washington, DC: Smithsonian Institution Press.
A collection of papers on the interplay of Hawaii's geological and biological histories.

Ziegler, Alan C. (2002). *Hawaiian Natural History, Ecology, and Evolution.* Honolulu: University of Hawaii Press.
A comprehensive guide to the natural history of Hawaii, from the tectonic origins of the islands to the impact of humans.

TEACHING EXERCISE

....

TRACING THE EVOLUTIONARY ORIGINS OF PICTURE-WINGED *DROSOPHILA* SPECIES

Teacher's Manual

In this investigation, students use actual genetic data from 18 species of Hawaiian drosophilid flies to construct an evolutionary diagram that depicts lines of descent for four of the species. They then draw on information about the ages of the Hawaiian islands to develop explanations that might account for the geographical distribution of the species. Finally, they analyze the data in greater detail to reexamine and elaborate on their explanations.

The exercise is designed for students in grades 9 through 12 who are already familiar with basic concepts in evolution and genetics. It can be adapted to occupy two, three, or four class periods.

This investigation provides an opportunity for students to gain experience with the ideas presented in *Evolution in Hawaii* and in *Teaching About Evolution and the Nature of Science* (National Academy of Sciences, 1998).

It consists of four sections:

- **Teacher's Manual**
- **Data Tables**
- **Student Reading**
- **Student Worksheet**

Components of the Investigation

The investigation is organized according to five activities involved in inquiry-based investigations: engagement, exploration, explanation, elaboration, and evaluation.

In the **engagement** activity, students read a background document describing the drosophilid flies of Hawaii, their courtship rituals, their descent from a common ancestral species, and the speciation processes that led to their current diversity. The reading concludes with several review questions that can provide the basis for written responses or a classroom discussion.

In the **exploration** activity, students complete a worksheet to gain experience with the use of genetic data to construct an evolutionary tree. They then analyze a subset of the data that evolutionary biologists have gathered in recent decades to determine the evolutionary relationships of Hawaiian drosophilid species. By working in small groups, students can acquire a common set of experiences that will help them share their ideas and conclusions. They also can acquire and model the information and skills they will need to analyze larger and more complex data sets.

In the explanation activity, students use the evolutionary tree they have constructed along with information about the history of the Hawaiian islands to develop hypotheses that could explain the current distribution of the four species being studied. By presenting their hypotheses in the form of narratives, they can demonstrate their understanding of important concepts in their own words and focus on understanding the ideas central to the investigation.

In the elaboration activity, students analyze additional data to decide whether their hypotheses need to be modified or whether to clarify and extend their explanations. The elaboration phase can be extended (bringing the total time spent on the exercise to three or four class periods) or minimized (in which case the exercise can be done in two class periods) depending on the time available. This activity also can lead to further investigations for interested students.

The evaluation activity is an ongoing diagnostic process that allows teachers to determine if learners have attained understanding. A rubric is provided to help assess student learning. Summative questions and assignments also are provided to evaluate the extent to which students understand the concepts developed in the investigation.

Alignment with the *National Science Education Standards*

According to the *National Science Education Standards* (National Research Council, 1996), science education needs to provide students with three kinds of scientific skills and understanding. Students need to learn the principles and concepts of science. They need to understand and be able to apply the skills and procedures of inquiry that scientists use to investigate the natural world. And they need to understand the nature of science as a particular kind of human endeavor.

This teaching exercise fosters learning in all three of these areas. With regard to the principles and concepts of science, it embodies the following understandings drawn from the *National Science Education Standards:*

- Species evolve over time. Evolution is the consequence of the interactions of (1) the potential for a species to increase its numbers, (2) the genetic variability of offspring due to mutation and recombination of genes, (3) a finite supply of the resources required for life, and (4) the ensuing selection by the environment of those offspring better able to survive and leave offspring.

- The great diversity of organisms is the result of more than 3.5 billion years of evolution that has filled every available niche with life forms.

- Natural selection and its evolutionary consequences provide a scientific explanation for the fossil record of ancient life forms, as well as for the striking molecular similarities observed among the diverse species of living organisms.

- The millions of different species of plants, animals, and microorganisms that live on earth today are related by descent from common ancestors.

- Biological classifications are based on how organisms are related. Organisms are classified into a hierarchy of groups and subgroups based on similarities that reflect their evolutionary relationships. The species is the most fundamental unit of classification.

This activity also provides all students with opportunities to develop abilities of scientific inquiry as described in the *National Science Education Standards*. Specifically, it enables them to:

- identify questions that can be answered through scientific investigations;

- use appropriate tools and techniques to analyze and interpret data;

- develop descriptions, explanations, predictions, and models using evidence;

- think critically and logically to make relationships between evidence and explanations;

- recognize and analyze alternative explanations and predictions; and

- communicate scientific procedures and explanations.

Finally, this activity provides all students with opportunities to develop understandings about inquiry and the nature of science. Specifically, it incorporates the following concepts:

- Different kinds of questions suggest different kinds of scientific investigations.

- Current scientific knowledge and understanding guide scientific investigations.

- Technology used to gather data enhances accuracy and allows scientists to analyze and quantify results of investigations.

- Scientific explanations emphasize evidence, have logically consistent arguments, and use scientific principles, models, and theories.

- Science distinguishes itself from other ways of knowing and from other bodies of knowledge through the use of empirical standards, logical arguments, and skepticism, as scientists strive for the best possible explanations about the natural world.

Background Information for Teachers

A remarkable implication of the theory of evolution is that all of the species that exist on the earth today are descended from common ancestors. In other words, if several species are compared, an evolutionary tree can be drawn that suggests how those species are related by common ancestry.

In this exercise, students gain familiarity with the idea of common ancestry and descent by investigating the evolutionary relationships of 18 species of drosophilid flies that live on the Hawaiian islands. Morphological and genetic data indicate that the approximately 800 species of drosophilid flies in the Hawaiian islands are descended from a single colonizing species that came to the islands millions of years ago. In studying the evolution of the Hawaiian drosophilids, biologists have focused on a group of about 100 species that can be grown in the laboratory and have large wings with distinctive black markings. Known as the picture-winged drosophilids, these species lay their eggs in rotting stems and bark and in bark moistened by the saps, or fluxes, exuded by trees. After the eggs hatch, the larvae feed on bacteria in those substances before pupating and becoming adult flies.

For the drosophilid flies of Hawaii, biologists have a particularly informative marker to trace their evolutionary relationships. During the larval stage of a fly's life, the salivary glands produce cells containing a special kind of giant chromosome known as a polytene chromosome. If extracted from the salivary cells and stained, these polytene chromosomes exhibit a hundred or more light and dark bands of varying sizes arrayed along their lengths.

On rare occasions a chromosome within an individual fly will undergo a particular kind of rearrangement. The chromosome breaks at two random points along its length. Usually these breaks are correctly repaired, so that the chromosome sequence is unchanged. But in some cases, the DNA breaks are repaired in a way that causes a section of DNA to be reversed in orientation. In that case, a portion of the chromosome, with its characteristic light and dark bands, appears to have been rotated 180 degrees in the chromosome. These rare chromosomal inversions can be of varying lengths, can occur on any chromosome, can occur within other inversions, and generally have no effect on the behavior or morphology of the fly. Because these inversions occur at random along the chromosomes and are of different sizes, each one is essentially unique.

If this kind of inversion mutation occurs in the sperm or egg cells of a particular fly, that fly can pass the rearranged chromosome to the next generation. Sometimes the inversion will become more and more common with each new generation, until an entire species of flies has the inversion. Therefore, if two species share an identical inversion, they must be descended from a common ancestral species that also had that inversion. If one species has an inversion that is *not* present in another species then that inversion must have occurred after the two species diverged from a common ancestor.

These inversions can be used to trace the evolution of many of the Hawaiian drosophilid species from common ancestral species. In Hawaii, populations of flies have speciated as they have adapted to new kinds of habitats or have acquired different mating behaviors. Speciation also has often followed founder events, when a single fertilized fly or several flies either traveled or were transported from one island to another island or between habitable but geographically isolated portions of the same island. Successful colonization is more likely when founders from older islands in Hawaii move to younger islands, since younger islands generally contain fewer competing species of drosophilids.

The Big Island of Hawaii is home to 26 distinct species of picture-winged *Drosophila* flies that have been found to live only on this island. The Big Island is the youngest of the Hawaiian islands, so these species likely formed since the volcanoes of the island emerged above the ocean and became vegetated less than half a million years ago. A major objective of this exercise is to explore the relationships between these "new" species and the species living on the older Hawaiian islands.

WHAT STUDENTS NEED TO KNOW

To perform this exercise, students will need a basic understanding of chromosome structure, banding patterns, and chromosomal inversions. They also will need to be familiar with the concept of biological species, the basic mechanisms of speciation, the relationship of geographic isolation to speciation, and the construction of phylogenetic trees. They can acquire some of this information from the Student Reading and Student Worksheet in this exercise. More information can be found in *Teaching About Evolution and the Nature of Science* (National Academy of Sciences, 1998).

Engage

In the engagement phase of the investigation, students are introduced to the biology and evolutionary history of the drosophilid flies of Hawaii through a reading and set of review questions.

Student Objectives

■ Learn about some aspects of the basic biology of drosophilid flies.

■ Recognize that the banding patterns of polytene chromosomes from individual flies, combined with information about the morphology, habitats, and behaviors of those flies, can be used in some cases to trace the evolutionary relationships of separate fly species.

■ Recognize that the unique history and geology of the Hawaiian islands contribute to the striking species diversity found there.

Materials

The materials provided are the Student Reading and Student Worksheet, a map of the major Hawaiian islands showing their ages, a diagram showing the evolutionary relationships among mammalian groups, and a photograph of a polytene chromosome.

Teaching Strategies

After students have completed the reading, teachers can ask them to answer the following questions, which also appear at the end of the reading. (These questions also can serve as a springboard to a classroom discussion.)

1. What characteristics of the Hawaiian islands might have led to the enormous diversity of drosophilid species in the Hawaiian islands? (Among the possible responses are the variety of habitats found in relatively close proximity on the islands, the separation of habitats within and among islands, and the lack of competing organisms. Each of these characteristics can be compared to conditions on a larger landmass.)

2. How might the diversity of mating behaviors be related to the diversity of drosophilid species in the Hawaiian islands? (Students could explore the idea that the evolution of new mating behaviors may contribute to speciation.)

3. What evidence suggests that all of the drosophilid species in the Hawaiian islands have descended from a single fertilized female fly that colonized the islands millions of years ago? (Students could discuss such evidence as similarities in shape, details of body form, behavior, chromosomal structure, and protein or DNA sequences.)

4. How could individual flies spread from island to island? (Students could suggest several possible mechanisms and ways of gauging their plausibility.)

Explore

In the exploration phase of the investigation, students use actual genetic data to trace evolutionary lines of descent for four species of Hawaiian drosophilid flies. By comparing their findings to information about the geology of the Hawaiian islands, they then conclude that new species tend to appear on younger islands.

Student Objectives

- Learn how to construct a phylogenetic tree showing the evolution of descendant species from a common ancestral species.

- Understand that chromosomal inversions can be used both to determine (1) that two species are descended from the same ancestral species and (2) to distinguish species descended from the same ancestral species.

- Relate the origins of species to the ages of the islands on which those species live.

Materials

Students will work from the data in Table 1. Later, they will compare their evolutionary trees to the dates shown on the map of the Hawaiian islands.

Teaching Strategies

The students' objective is to derive evolutionary pathways from the data in Table 1 for four of the species described in that table: *Drosophila heteroneura, D. hanaulae, D. substenoptera,* and *D. primaeva.* Students should work in small groups so they can arrive at a consensus about the most reasonable structure of the tree. They should construct a single tree agreed upon by all members of the group so they gain experience with achieving consensus based on empirical evidence and defending their reasoning.

Table 1 begins by listing 11 chromosomal inversions found in the species *D. heteroneura.* Following this are the inversions found in 12 other species of picture-winged *Drosophila* found in the islands. These are incomplete listings of the inversions, since only the 11 inversions found in *D. heteroneura* are given in this table. Most picture-winged species have some inversions not found in *D. heteroneura.* For example, *D. setosimentum,* which also is found on the Big Island of Hawaii, has a total of 21 inversions, as shown on the top line of Table 2. However, it shares only four of its inversions with *D. heteroneura,* as shown on the bottom line of Table 1. Note, too, that some species have identical inversions so they need to be distinguished using other physical or behavioral characteristics.

Students can perform this exercise without exploring all of the dimensions of the data contained in Tables 1 and 2. At the same time, advanced students can use the data provided in Tables 1 and 2 to extend the exercise in many productive directions.

Once the class has used the data in Table 1 to reconstruct the evolutionary relationships among *D. heteroneura, D. hanaulae, D. substenoptera,* and *D. primaeva,* they should be able to conclude, by comparing their trees with the geological ages of the islands, that new species tend to appear on younger islands.

Explain

In the explanation phase of the investigation, students should describe to each other and to the class the conclusions drawn from the data. They then should construct plausible explanations that can account for the observed data. In developing these explanations, students should explore the limitations of the data in providing complete explanations and consider what additional kinds of evidence might be gathered to support their hypotheses.

Student Objectives

- Use the evolutionary relationships developed by their groups and information about the geology of the Hawaiian islands to construct a historical narrative that could explain the current distribution of these four species.

- Present their narratives in small groups to the class or to the teacher.

- Analyze which parts of their narratives the evidence supports and which parts the evidence does not support.

Materials

Students will continue to rely on the Student Reading, the map of the Hawaiian islands, and the data in Table 1 to provide the raw material for their narratives.

Teaching Strategies

Constructing a narrative gives students the opportunity to relate easily understood events—such as the journey of an individual fly from one island to another—to data that may seem to have little relevance to the everyday world. Such accounts can make abstract concepts such as speciation concrete and accessible. Putting the conclusions they draw from the data in the form of a story allows students to demonstrate their understanding of important concepts in their own words, emphasizing the understanding of the idea rather than their knowledge of facts or terminology.

To initiate this part of the investigation, teachers could assign the following task: *Write a narrative describing how the four species of picture-winged flies may have evolved from a common ancestor.*

The entire class, working together, could write the narrative, with the teacher guiding the class in selecting conclusions consistent with the data. Or the teacher and class could work together to begin the story, with students finishing the story working in groups. Alternately, the students could write the stories individually or in small groups.

One way to begin the writing of narratives would be to provide students with all or part of the following sample narrative:

About four million years ago on the island of Kauai there lived an interbreeding population of picture-winged flies. Every member of this population had the four chromosomal inversions i, k, o, and b. About three million years ago a storm on Kauai resulted in a large number of flies being swept off the island and carried over the sea. Most of the flies perished. By chance, one or several fertilized female flies could have landed on the more recently formed island of Oahu. Also by chance, some of the descendants of these fertilized females eventually acquired several other chromosomal inversions, the ones labeled p, q, s, and d in Table 1. Over many generations these inversions became more common until all of the members of the Oahu population of flies carried them. . . .

The narrative should address these points:

1. What are the different ways in which the flies could have spread from island to island?

2. Of the 11 inversions shown in the table, which are likely to be older and which younger, both for the four species examined and for all 13 of the species in the table?

3. Do new species tend to form on younger islands?

After the narrative or narratives have been written and shared, the following questions could help guide a class discussion:

1. Which aspects of the narrative are supported by the limited set of data available?

2. Which aspects of the narrative are not supported by the data?

3. What additional data can be collected and analyzed to modify, verify, or augment the narrative?

Elaborate

In the elaboration phase of the investigation, students can deepen and enrich their understanding of the evolution of the drosophilid flies of Hawaii by analyzing additional data and by modifying their narratives to take those data into account. This phase of the investigation is open-ended, in that students can do as much as they wish with the additional data. Typically, completing this phase will require devoting a third or fourth class period to the exercise.

Student Objectives

- Recognize that hypotheses often must be modified in the light of new data.

- Recognize that a particular data set does not necessarily support an explanation.

- Identify what additional data are needed to extend the explanation.

Materials Needed

Students will continue to work with the data in Table 1 and with the map of the Hawaiian islands. They also can refer to the data in Table 2.

Teaching Strategies

One way to demonstrate to students how their narratives need to be modified in light of new data is to lead a guided class discussion, using these questions:

1. Based on the data included in Table 1, how could the species *D. setosimentum* be added to the evolutionary tree? (As shown in Table 1, *D. setosimentum* has the same inversions as does *D. primaeva*, but because *D. setosimentum* lives on Hawaii and *D. primaeva* lives on the older island of Kauai, a logical hypothesis is that *D. setosimentum* originated through a founder event.)

2. What additional data could be used to test this hypothesis, and what data would support it? (For example, would additional inversions found in *D. setosimentum* that are not found in *D. primaeva* support this hypothesis?)

3. Does each species always have a new and different set of inversions?

Another possible elaboration would be to add to the earlier phylogenetic tree the species *D. silvestris,* which is also represented in Table 1. The following questions could lead students to examine these data:

1. Where does *D. silvestris* fit into the evolutionary tree relative to the other four species? (Since it has same inversions and is found on the youngest island, the inversion data indicate that it may have evolved relatively recently.)

2. Is it possible to determine whether *D. heteroneura* or *D. silvestris* is the more recently evolved species? (On the basis of the data in Table 1, it is not possible.)

3. How are *D. silvestris* and *D. heteroneura* related to *D. setosimentum?* How do you know? (Note that in Table 2, *D. setosimentum* has the most inversions and *D. heteroneura* the fewest, the opposite of their positions in Table 1.)

Another possible extension of this investigation makes use of the data in Table 2. This table contains inversion data for 10 species of drosophilids, beginning with *D. setosimentum.* Five of the 10 species in Table 2 are represented in Table 1 as well, while five of the species are different.

A possible task is to draw an evolutionary tree for all 10 of the species in Table 2 using the methods developed for the previous data set. This tree then can be compared to the ages of the islands where the species live today.

In Table 2, the species *D. heteroneura* has the fewest inversions, whereas in the previous data set it had the most. Similarly, *D. setosimentum* had the fewest inversions in the first data set but has the most in the second. How can these results be reconciled?

Questions that can guide this extension of the elaboration phase include:

1. Which groups of fly species have identical sets of inversions, and which flies have unique sets?

2. What is a good way to express numerically the similarities and/or differences among these species based on the inversion data?

3. Given the data available to you, which island was the most likely home for the ancestral species of these flies?

4. What evidence do the tables contain to support the idea that new species tend to appear on younger islands?

Evaluate

According to the *National Science Education Standards:*

[Students need] to participate in the evaluation of scientific knowledge. . . . What data do we focus on first? What additional data do we consider? What patterns exist in the data? Are these patterns appropriate for this inquiry? What explanations account for the patterns? Is one explanation better than another? In supporting their explanations, students have drawn on evidence to derive a scientific claim. Students have assessed both the strengths and weaknesses of their claims.

Questions that could be used to assess students' understanding after the investigation has been completed include the following:

1. Which data are important in establishing the evolutionary relationship between *D. hanaulae* and *D. oahuensis* (which have the same inversions but live on different islands)?

2. Which data are important in establishing the evolutionary relationship between *D. hanaulae* and *D. obscuripes* (which live on the same island but differ by one inversion)?

Students should be able to describe the methods they used to derive the phylogenetic trees generated during this exercise, how the data were analyzed, and what conclusions or generalizations can be drawn from the data. An advanced level of understanding would involve the ability to describe how the same species can appear in both data tables with quite different inversion data.

Students' work throughout this investigation also can be evaluated on an ongoing basis through the use of a rubric such as the one on the following page. This rubric has been modified from one provided by the International Baccalaureate Organization (IBO) for the assessment of student work in experimental science courses.

Rubric

	WORKING ON A TEAM	RECOGNIZING CONTRIBUTIONS OF OTHERS	EXCHANGING AND INTEGRATING IDEAS	APPROACHING SCIENTIFIC INVESTIGATIONS	WORKING IN AN ETHICAL MANNER
COMPLETE	Collaborates with others in order to complete the task.	Expects, actively seeks and recognizes the views of others.	Exchanges ideas with others, integrating them into the task.	Approaches the investigation with self-motivation **and** follows it through to completion.	Pays considerable attention to the authenticity of their explanations by working with their own group to develop original explanations.
PARTIAL	Requires guidance to collaborate with others.	Acknowledges some views.	Exchanges ideas with others but requires guidance in integrating them into the task.	Approaches the investigation with self-motivation **or** follows it through to completion.	Pays some attention to the authenticity of their explanations, but copies some ideas from others.
NOT AT ALL	Is unsuccessful when working with others.	Disregards views of others.	Does not contribute.	Lacks perseverance and motivation.	Pays little attention to authenticity of their work.

Data Tables

Table 1 Species That Share One or More of the 11 Inversions Found in
D. heteroneura of the Big Island of Hawaii

SPECIES	CH X	CH#3	CH#4	ISLAND
*D. heteroneura**	i j k o p q r s t	d	b	Hawaii
D. silvestris	i j k o p q r s t	d	b	Hawaii
D. planitibia	i j k o p q r s t	d	b	Maui
D. differens	i j k o p q r s t	d	b	Molokai
*D. hanaulae**	i j k o p q - s t	d	b	Maui
D. obscuripes	i j k o p q - s -	d	b	Maui
D. hemipeza	l j k o p q - s t	d	b	Oahu
D. oahuensis	i j k o p q - s t	d	b	Oahu
*D. substenoptera**	i j k o p q - s -	d	b	Oahu
*D. primaeva**	i - k o - - - - -	-	b	Kauai
D. ornata	i - k o - - - - -	-	b	Kauai
D. setosifrons	i j k o - - - - -	d	b	Hawaii
D. setosimentum	i - k o - - - - -	-	b	Hawaii

NOTE: Each inversion is denoted by a letter. A dash denotes that the species does not have that inversion. Nine of the inversions studied in this data set have occurred on the X chromosome, and one each has occurred on chromosome 3 and on chromosome 4. CH = Chromosome.

*The evolutionary relationships of these four species are explored in detail in the exploration phase of the exercise.

Table 2 Species That Share One or More of the 21 Inversions Found in
D. setosimentum of the Big Island of Hawaii

SPECIES	CH X	CH#2	CH#3	CH#4	CH#5	ISLAND
D. setosimentum	ikouvwxym^2	cd	fjkl	bopqb^2	f	Hawaii
D. ochrobasis	ikouvwxym^2	cd	fjkl	bopqb^2	f	Hawaii
D. adiastola	ikouvwxy-	cd	fjk-	bopq-	f	Maui
D. hamifera	ikouvwxy-	cd	f-k-	bopq-	f	Maui
D. toxochaeta	ikouvwxy-	cd	fjk-	bopq-	f	Molokai
D. touchardia	ikouvwxy-	cd	fjk-	bopq-	f	Oahu
D. ornata	ikou--xy-	cd	-jk-	bo---	-	Kauai
D. primaeva	iko------	c-	----	b----	-	Kauai
D. setosifrons	iko------	--	----	b----	-	Hawaii
D. heteroneura	iko-----	--	----	b----	-	Hawaii

NOTE: Inversions m^2 and b^2 are verbalized as "m-two" and "b-two."
CH = Chromosome.

Student Reading

You wouldn't think that biologists could learn much about evolution by studying the flies that live on the hills and mountainsides of the Hawaiian islands. But these are not ordinary flies.

Approximately 800 species of flies belonging to the genera *Drosophila* and *Scaptomyza* live in the forests of Hawaii. Collectively known as drosophilids, these flies differ significantly from the houseflies that buzz around kitchens and garbage piles. Houseflies belong to a different biological family and were brought to Hawaii by human colonizers over the last few hundred years. The drosophilids have lived in Hawaii for millions of years and are the products of a spectacular evolutionary history.

The different drosophilid species in Hawaii vary greatly. Some species are large and others are small. They have differently shaped bodies and contrasting behaviors. Different species lay their eggs in leaves, bark, fungi, fruit, or even spider eggs. One particularly well-studied group of about a hundred species has bold black markings on their wings. These species are known as picture-winged drosophilids.

Many drosophilid species have elaborate sex lives. First the males establish a mating territory called a lek. Then they defend this area from other males of the same species. Males of one species, called *Drosophila heteroneura,* use their hammer-shaped head as a battering ram to drive other males away (see Figure 1). The males of other species lock legs and wings and wrestle each other into submission until one flees. In another species, males make a buzzing roar with special muscles in their abdomen.

When a female fly visits the lek, the male gets to work. In many species, the males have an elaborate but specific dance that they use to attract females. Others buzz their wings in a special way or place their heads under the female's wings. One species releases attractant chemicals known as pheromones in the female's direction.

Even if the male does everything perfectly, his efforts may not pay off. If the female chooses not to mate with that male, she will fly away.

Figure 1

D. heteroneura males use their hammer-shaped heads to defend their territories. (Photograph courtesy of Kevin Kaneshiro.)

The drosophilid species of Hawaii are so different from each other that it is hard to believe they are all descended from a common ancestral species. But the biological evidence is compelling. All of the approximately 800 species of drosophilids on Hawaii are descended from the members of a single species of flies that made their way to the Hawaiian islands many millions of years ago. In fact, all of the Hawaiian drosophilids may be descended from a *single* fertilized female that reached the Hawaiian islands!

No one will ever know exactly how this fertilized fly traveled to the islands (if it was just one). Maybe it was blown to Hawaii in a storm. Maybe it was carried on a solid object to the islands. But somehow it survived the trip, and when it arrived at the islands it began to produce offspring.

Biologists suspected for a long time that the Hawaiian drosophilids were descended from a single ancestral species because of physical characteristics they share. But one piece of evidence has been particularly persuasive. When drosophilids are in their larval stage, biologists can examine cells from their very large salivary glands. These cells contain a special kind of chromosome called a polytene chromosome. In a polytene chromosome, many copies of a single DNA molecule line up side by side, making the chromosome large enough to see with a microscope. If a stain is added to the salivary tissue, the chromosomes exhibit a characteristic pattern of hundreds of light and dark bands along their lengths. The polytene chromosomes of the Hawaiian drosophilids, along with other genetic data, point toward a common origin for these very different species of flies.

Polytene chromosomes also can be used to trace the evolutionary history of individual drosophilid species. These chromosomes can reveal a particular kind of mutation called an inversion. Inversions are rare events that result in a section of a chromosome becoming reversed with respect to the original sequence. Inversions occur as a consequence of molecular damage and repair processes in the nucleus of the cell. Biologists can recognize these inversions in polytene chromosomes because the pattern of light and dark bands along part of the chromosome is reversed (see Figure 2).

These inversions provide a way of reconstructing evolutionary relationships. For example, if two species of picture-winged drosophilids both have the same inversion, they must be descended from an ancestral species that had that inversion. If one species has an inversion that the other does not, that inversion must have occurred in the time since the two species split from a common ancestor.

For example, think about the original fertilized fly that may have been the ancestor of all the Hawaiian drosophilids. As that fly began to produce offspring, they would have closely resembled the original colonist. Over time, this small population of flies would grow.

Eventually a group of flies must have become separated in some way from the others, either geographically, behaviorally, or ecologically. Females from this isolated population might have laid their eggs in different substances or become adapted to a different habitat. The members of this population

Figure 2

Polytene chromosomes from larvae of drosophilid flies provide a way to detect chromosomal inversions. In the chromosome shown here, the chromosomal segment between the two indicated points is inverted compared to the same chromosome in other species. (Photograph courtesy of Hampton L. Carson, based on an original photograph by Harrison D. Stalker, Washington University, St. Louis.)

might have developed new mating behaviors. At the same time, this population could acquire chromosomal inversions that would distinguish it from the original population of flies.

At some point, the daughter population would have diverged so greatly from the original population that its members generally would not recognize courtship signals or body markings from the original population. As a result, individuals from the isolated population would no longer mate and produce offspring with flies from the original population. If, as time progressed, physical and behavioral traits became sufficiently different between the two populations so that individuals from the two populations could no longer produce viable or fertile offspring, the daughter population would be considered a new species.

This speciation process has occurred many, many times among the drosophilid flies of Hawaii. Sometimes it happened when a small group of flies, or just a fertilized female, was carried, was blown, or flew from one part of an island to another. Other times it occurred when a fly journeyed or was transported from one island to another. Because of their volcanic origins, the Hawaiian islands have different ages, with the younger islands to the southeast. As each new island rose from the waves, small populations of flies or single fertilized females made their way from older islands to newer ones, where their descendants could become increasingly distinct from the ancestral species.

The Hawaiian islands also have many kinds of environments, from lowland rainforests to dry upland forests. Different drosophilid species have evolved adaptations that enable them to thrive in different types of forests.

Also, on the mainland, a new drosophilid species would face competition from other insects already living in an area or from predators. But because the Hawaiian islands are isolated in the middle of the Pacific, new species had far fewer competitors or predators when they appeared. On their island paradise, the drosophilids have flourished.

Biologists call the evolution of many new species from a single ancestral species an adaptive radiation. The adaptive radiation of the Hawaiian drosophilids is one of the most dramatic examples of evolution found anywhere in the world.

Review Questions

1. What characteristics of the Hawaiian islands might have led to the enormous diversity of drosophilid species in the Hawaiian islands?

2. How might the diversity of mating behaviors be related to the diversity of drosophilid species in the Hawaiian islands?

3. What evidence suggests that all of the drosophilid species in the Hawaiian islands have descended from a single fertilized female fly that colonized the islands millions of years ago?

4. How could individual flies spread from island to island?

Student Worksheet

In this investigation of evolution in Hawaii, you will use actual genetic data from 18 species of Hawaiian drosophilid flies to construct an evolutionary tree that depicts evolutionary lines of descent for four of those species. These species belong to a group of about a hundred well-studied species known as the picture-winged drosophilids that have prominent black markings on their wings. Once the tree is complete, you will use it along with information about the geology of the Hawaiian islands to propose a series of events that could explain the current distribution of the four species.

Materials Needed

■ Student Reading

■ Data Tables summarizing chromosomal inversions in 18 species of picture-winged drosophilids

Background

First, you need to gain experience drawing an evolutionary tree and relating that diagram to specific genetic events. As shown in Figure 1, phylogenetic trees depict the evolution of two or more descendant species from a single ancestral species, with lines connecting the ancestral species to the descendant species. In the drosophilid flies of Hawaii, these descendant

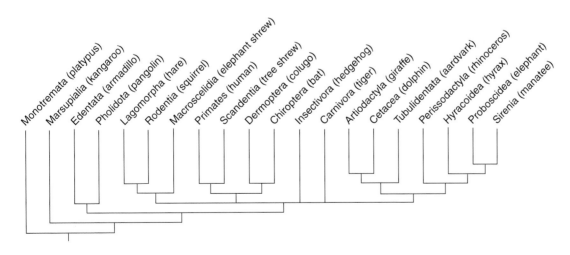

Figure 1

An evolutionary tree of the living groups of mammals demonstrates their relationships, though some of the details of the tree remain controversial or ambiguous. A representative member of each group is shown in parentheses after the group name. In addition to the groups shown here, other mammalian groups have gone extinct. (Diagram adapted from Colin Tudge, *The Variety of Life: A Survey and Celebration of All the Creatures That Have Ever Lived.* New York: Oxford University Press, 2000.)

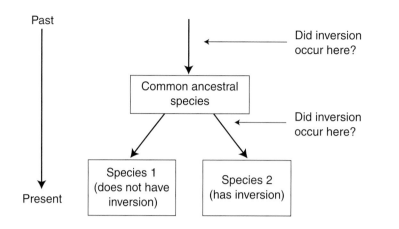

Past

Present

Did inversion occur here?

Did inversion occur here?

Common ancestral species

Species 1 (does not have inversion)

Species 2 (has inversion)

Figure 2

If two species descended from a common ancestral species differ by the presence of a chromosomal inversion, when must the inversion have occurred?

species often can be distinguished by the presence or absence of specific chromosomal inversions.

In Figure 2, one species descended from a common ancestral species has an inversion that the other does not. Where on this diagram must have the inversion have occurred?

This is the kind of analysis you will use to construct a more detailed evolutionary tree in this investigation.

Main Activity

In this part of the investigation, you will use the chromosomal inversion data from Table 1 to construct an evolutionary tree for the following four species: *Drosophila heteroneura,*

D. hanaulae, D. substenoptera, and *D. primaeva.* You should indicate on the tree where the inversions occurred that can be used to determine the species' evolutionary relationships.

Once the diagram is complete, compare the evolutionary relationships of the four species with the ages of the Hawaiian island on which they live (see Figure 3). What conclusions can you draw about how new species of flies appear in Hawaii?

The final part of this main activity is to construct a narrative account of how the four species of drosophilid flies might have evolved over time. Your account should address these points:

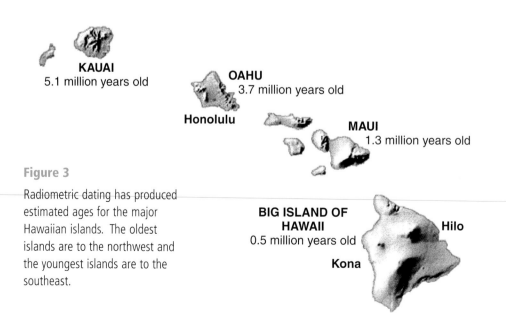

KAUAI
5.1 million years old

OAHU
3.7 million years old

Honolulu

MAUI
1.3 million years old

Figure 3

Radiometric dating has produced estimated ages for the major Hawaiian islands. The oldest islands are to the northwest and the youngest islands are to the southeast.

BIG ISLAND OF HAWAII
0.5 million years old

Hilo

Kona

1. How might the flies have traveled from one island to another?

2. How might new species of flies have evolved on the islands where they now live?

3. In what order did the inversions shown in the table occur?

4. Which species might be older and which might be younger?

Elaboration

In the elaboration phase of the investigation, additional data from Table 1 or the data from Table 2 can be analyzed. Using all the data in Table 1 permits the construction of a much larger phylogenetic tree showing the evolutionary relationships among the 13 species of flies described in the table. The data in Table 2 permit the construction of a second tree that reveals an even more complex set of evolutionary relationships among the 18 species of picture-winged flies represented in the two tables.

Questions to consider in light of these additional data include the following:

1. Are the chromosomal inversion data always sufficient to distinguish species?

2. Does the presence of an inversion necessarily mean that one species is younger than another?

3. Can a relatively young species occur on an older island?

4. What are the mechanisms that would enable this to occur?